THE WORLD OF PARTICLES

I0491080

INDIA · SINGAPORE · MALAYSIA

Notion Press

No.8, 3rd Cross Street, CIT Colony,
Mylapore, Chennai,
Tamil Nadu – 600004

First Published by Notion Press 2020
Copyright © Pradyumn 2020
All Rights Reserved.

ISBN 978-1-64899-949-9

THE WORLD OF PARTICLES

UNRAVELING THE RECIPE
TO BUILD OUR UNIVERSE

PRADYUMN

INDIA · SINGAPORE · MALAYSIA

**INDIC** ACADEMY

INDIC PLEDGE

- *I celebrate our civilisational identity, continuity & legacy in thought, word and deed.*

- *I believe our indigenous thought has solutions for the global challenges of health, happiness, peace, and sustainability.*

- *I shall seek to preserve, protect and promote this heritage in doing so,*
    - *discover, nurture and harness my potential,*
    - *connect, cooperate and collaborate with fellow seekers,*
    - *be inclusive and respectful of diverse opinions.*

ABOUT INDIC ACADEMY

Indic Academy is a non-traditional 'university' for traditional knowledge. We seek to bring about a global renaissance based on Indic civilizational and indigenous thought. We are pursuing a multidimensional strategy across time, space and cause by establishing centers of excellence, transforming intellectuals and building an ecosystem.

Indic Academy is pleased to support this book.

Contents

Preface

Particle Physics has always been my favourite domain to explore. During my high school years, I was obsessed with Particle Physics. So much so, that I started building my miniature particle accelerator as my first-year high school project. Unfortunately, I was unable to arrange 20KV for building a particle generator and the project turned out to be a theoretical one, rather being experimental. Based on my then understanding of the microscopic world, I started writing about the quest of finding the building blocks of our universe. Since then, it's been four years that I am working on this book. It has been an amazing journey through the world of particles!

In this book, I have described the building blocks of our universe and the associated theories. The book is written for readers who would like to explore the world of particles and understand the ultimate theory of the universe with minimal mathematical understanding. Therefore, the book welcomes technical as well as non-technical readers to enjoy and explore the world of particles.

The book begins with the introduction of the atom by Dalton in 1808 and follows-up with the discoveries of other particles like electron, proton, and neutron. It describes the quest among the scientists to understand the shape of the atom, nature of the then discovered particles and the newly developed quantum mechanics.

It then proceeds towards the development of particle accelerators to unveil new Physics. It describes the working of the Large Hadron Collider (L.H.C.), a 27km circumference massive machine that creates extraordinary levels of energy to discover new physics. It sets a bar for the readers to understand the contributions of scientists across the globe, to understand the building blocks of our universe and to develop the ultimate theory of everything that will describe all known laws of our universe into a single theory.

It also describes the discoveries made by the L.H.C. and theoretical and experimental efforts made by the scientists to develop a Standard Model of Particle Physics (SMP). The model enhances the reader's understanding of the elementary particles and fundamental forces that make-up the recipe to build our universe. It then describes the contemporary efforts by the scientists to unveil Physics beyond the Standard Model.

The book provides a list of particles that have been discovered and hypothesized in the past sixty years and aims to give a gist of the particles in Particle Physics. It then describes the Classical Field Theory and Quantum Field Theory (QFT) which are the modern developments in Particle Physics. It builds a strong base of various field theories like Quantum Electrodynamics (QED), Quantum Chromodynamics (QCD), Quantum Flavordynamics (QFD) and Quantum Gravity. It introduces the reader with Feynman diagrams, CPT-symmetry, Higgs mechanism, String Theory, Loop Quantum Gravity and helps to build a foundation to understand the unification of forces and the ultimate theory of the Universe i.e. the Theory of Everything.

About the Author

Teenage author of '*Journey Through the Dark Monster*', a primer on black holes and general relativity, Pradyumn is simply fascinated by Physics.

Pradyumn started his journey into Physics at the age of fifteen and became curious about the laws of the Nature. Eventually, he fell in love with Particle Physics for obvious reasons and started exploring the microscopic world. Along with this, he started a blog titled 'Physics Mindboggler' to share his learnings with others. The main agenda of Physics Mindboggler is to inspire, in addition to throwing light on various scientific aspects without confounding the reader.

Pradyumn firmly believes that research and development of technologies make up a major chunk of the economic growth of a country. Therefore, this field needs to be taken at the next level by today's young generation by thinking about the technological revolutions they can bring about, via research.

Pradyumn has also published research papers on '*Performance Analysis of Thermoelectric Generator by using Lead Telluride, Perovskites, Skutterudites and Tetrahedrites*' which was selected in WEENTECH (World Energy and Environment Technology Ltd.) International Conference on Energy, Environment, and Economics (ICEEE 2019) at

Heriot-Watt University, United Kingdom and *'Optimization and Analysis of Novel Thermoelectric Module'* which was selected in JVE (Journal of Vibroengineering) International Conference on Dynamics, Noise, Vibration and Smart Materials. With this and upcoming research works, Pradyumn is on his way to achieving his dreams.

He loves challenges and equally enjoys stress, as both test his capabilities. Setting everyday goals, living in the present and managing time polish him every day. Fully aware that all work and no play make Jack a dull boy, Pradyumn is also a Second Dan Black Belt in Karate Budokan International.

Acknowledgments

Unraveling the recipe of our universe was one of the most thrilling experiences I ever had. I never imagined that a tiny high school project could turn out into a book and inspire me to start my blog 'Physics Mindboggler'. It's been four years since then and I'm pleased that I finally made it through! However, this achievement is not entirely mine; it's because of many well-wishers and supporters who made this book possible. I would like to acknowledge all of them for their support and blessings.

The genesis of this book can clearly be traced back to my teachers at the St. Pius X High School and Vani Vidyalaya Jr. College, who inspired me to explore the realm of science and technology. The person who ignited the passion in me to raise fundamental questions and to seek answers to them through observation and experimentation was Prof. Shiv Yadav, my Physics teacher at Akhil Tutorials. In fact, this book is a result of the passion ignited by him.

All my teachers since then, right up to my current teachers, HoD, Dean and Vice-Chancellor at Bennett University, have only fuelled that fire. It would be as futile to even start thanking all these gurus, as it would be to thank my beloved parents and family. No words of gratitude can do justice to their contribution in making me who I am.

Writing this book was one thing, but getting it reviewed was quite another. A renowned self-taught Hungarian physicist Viktor T. Toth has been extremely kind in tirelessly and uncomplainingly reviewing theories/concepts covered in the book. I have also made extensive use of You Tube videos of the famous American Physicist, author, and science communicator Dr. Don Lincoln who is also a Senior Scientist at Fermilab with expertise in experimental Particle Physics. His Fermilab YouTube videos have played a crucial role in completion of this book. I have no words to thank them enough.

I express special gratitude towards my mentors at Bennett Hatchery Dr. Vinod Shastri, Dr. Abhinav Chaturvedi and Mr. Manish Mathur. The book would definitely have been written, but without their constant encouragement and support, it probably would not have been published so soon. I also thank Prof. Shajan Kumar of the Times School of Media for his expert guidance as well as for personally shooting a promotional video for the crowdfunding campaign for this book.

I thank Ms. Labdhi Gadhaiya and Ms. Swetha Muthukrishnan of Notionpress for their guidance and support in publishing this book and taking it to discerning readers like you.

Last, but not the least, I express special gratitude towards all my well-wishers who financially supported the publication of this book. My Bennett University mates contributed in overwhelming numbers through a 'Fund-a-Friend' campaign that was blessed by Ms. Revati Jain and

Dr. R. K. Shevgaonkar, initiated by Dr. Milind Padalkar and Ms. Manisha Shukla and supported by SPARK, the student Entrepreneurship Cell. I am equally thankful to numerous other Milaap contributors, who insisted on remaining anonymous.

– Pradyumn

List of Financial Supporters

1. Abhijan Ganguly
2. Abhinav Singh
3. Akash
4. Akhil Parim
5. Ambika
6. Ambuje Gupta
7. Aneesh
8. Anil
9. Anirban
10. Archana Deodhar
11. Arjun
12. Arvind Mishra
13. Avinash Purandare
14. Ayush Arora
15. Devendra Deshmukh
16. Dhriti Gaur
17. Garima
18. Hitesh Nair
19. Ishita Agarwal
20. Jagadiswari Panduri
21. Jhalak
22. Kabier
23. Kiran
24. Manasi Sadhale
25. Manav Hada
26. Mandar Wadadekar
27. Mayank Verma
28. Mayuresh Buchake
29. Mehul Darooka
30. Mohammed Adil
31. Mukesh
32. Nandita Mishra
33. Navya Kapahi
34. Nihil Gautam
35. Nikhil Saraswat
36. Parag Mahajan

37. Phalak Saifi
38. Pradyumn Jain
39. Radhika
40. Rajiv Ranjan
41. Raunak Gulati
42. Reeya Sharma
43. Riya
44. Rohan Patil
45. Ronald Fernandez
46. Runal Dahiwade
47. Sapan Shrimal
48. Shajan Kumar
49. Shivangi Mishra
50. Shreyans Jain
51. Shruti Sinha
52. Siddhart Patni
53. Siva Sanagala
54. Srijana Singh
55. Sudha Bhamidipati
56. Sujaya Rao
57. Sumit Roy
58. Sunil Parab
59. Tanvi Agarkar
60. Tarussi Singh
61. Uzma Hussain
62. Vaishali Sharma
63. Valay Jain
64. Varun Tejwani
65. Yash
66. Yuvraj

Introduction and Discoveries

When massive objects are quantized, things work very differently. We see very exciting phenomena which are very, very hard to see in macroscopic objects. In 1808, Dalton introduced us to the atom, which was the smallest indivisible particle of matter at that time. Various elements were discovered later on. Mendeleev was first who successfully arranged all known elements of that time in order. Later, in 1896, J.J. Thomson discovered a negatively charged particle which was named as an electron by H.A. Lorenz and therefore proved that atom is divisible. The research was done forward to discover its mass, charge, e/m ratio, etc. In the same year, Goldstein discovered proton, which was 1837 times the mass of the electron. Further research was continued and it leads to the discovery of the neutron by Chadwick in 1932. The discovery of these sub-atomic particles showed that the atom is no more indivisible.

But this was still not enough to understand the atom. After discovering electron, proton, and neutron, scientists step forward to know the shape of the atom. Among many scientists, Rutherford put forward an experiment of Alpha-scattering and showed that the entire mass of the atom is concentrated at its center. His model resembled the solar system, in which planets revolve around the sun. But still, this was not enough to resolve our thirst for the micro-world. Albert Einstein, Niels Bohr, Louis de Broglie, Erwin Schrodinger, Plank, and Paul Dirac discovered mathematical expressions of quantum theory which made a positive development in quantum mechanics.

Quantum Mechanics showed that the atom was not like we imagined. It didn't have orbits; instead, it has orbitals, where the maximum possibility of finding electrons can be mathematically calculated. So far scientists have discovered four orbitals that have their independent shapes and capacities. For more convenience, Quantum Mechanics can be defined as, "the branch of mechanics that deals with the mathematical description of motion and interaction of sub-atomic particles, incorporating the concepts of quantization of energy, wave-particle duality, the uncertainty principle, and corresponding principle."

Large Hadron Collider (LHC)

Particle physics, high energy physics, and quantum physics are some of my fascinating fields in physics. The mysterious quantum realm never fails to surprise scientists and particle accelerators are the technology that could help them to discover new physics. One of many particle accelerators is the CERN's large hadron collider (L.H.C.), a 27km circumference massive machine that creates extra-ordinary levels of energy resulting to create micro black holes. This gigantic machine discovered Higgs boson (famously called the 'God particle') in 2012 about which I will be talking in upcoming chapters. In a nutshell, Higgs boson is a particle which confirms the existence of the Higgs field, with whom other particles interact via Higgs mechanism (Chapter 6) to gain mass.

Since 1920's particle accelerators are being made. These machines took us one step further. They discovered new sub-atomic particles like muon, tau, sigma, pions, etc. Presently among 30,000 particle accelerators The Large Hadron Collider (LHC) at Geneva, the Switz-France border is the largest. All scientists across the globe come here and recreate the moment, present in the universe less than a billionth of a second after Big Bang. This massive machine with 27 km circumference is the treasure land for hunting many subatomic particles. It has given zoo of particles (Chapter 5). The big discovery made by LHC is the Higgs Boson, whose field contributes mass to sub-atomic particles. Detailed information on this will be done later, but before that let us see the basic functions of LHC.

STAGE 1: PARTICLE GENERATION

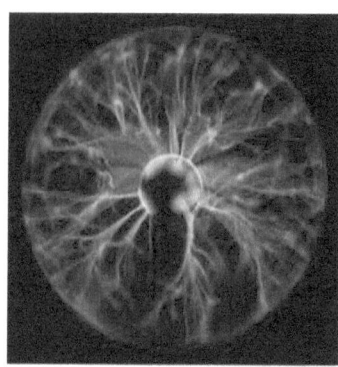

Figure 2.1: Plasma ball

A cylinder of compressed hydrogen gas is the beginning of the world's largest and powerful particle accelerator. Hydrogen gas is sent in the source chamber of a linear accelerator for protons (Linac 2) where its electrons are stripped off leaving behind the protons of each hydrogen gas. This stripping off electrons is done simply via plasma.

STAGE 2: PROTON SYNCHROTRON BOOSTER

The proton is simply guided via electric field and superconductive massive magnets causing to keep the particles in focus i.e. at the center. Under ultra-high vacuum, proton enters the Proton Synchrotron Booster (PSB). In this 157m circumference booster, the particles are accelerated to 91.6% of the speed of light.

STAGE 3: PROTON SYNCHROTRON

Constructing linear accelerators is impractical, therefore, scientists and engineers use synchrotron (a circular path) which does an efficient work in less space. From PSB the particles are sent to Proton Synchrotron (PS). PS takes protons from PSB at a kinetic energy of 14GeV and lead (Pb) ions (coming from Low Energy Ion Ring (LEIR)) of 72MeV and converts them to 28GeV. This 628.3m path accelerates the proton to 99.9% of the speed of light.

It's a versatile machine that has accelerated protons, antiprotons, electrons, positrons and species of ions like lead (Pb). It's here where the point of transition is reached, a point where the energy added to the protons cannot translate into increased velocity as they are already approaching the limiting speed of light. Instead, the added energy manifests itself as an increasing mass of the protons. In PS the protons get 28 times heavier and are then channelled to 7km Super Proton Synchrotron (SPS) which can increase the energy up to 450GeV.

STAGE 4: INTO THE LARGE HADRON COLLIDER

Now the particles are ready to enter the massive accelerating ring of the 27km circumference. Here the proton beams are divided and accelerated in two different adjacent tubes in a clockwise and anticlockwise direction in each tube respectively.

The proton achieves the energy of 13TeV and velocity of about 99.999999% of the speed of light. Here the proton beams are guided with a strong magnetic field maintained by superconducting magnets in ultra-high vacuum.

MAGNETS USED IN L.H.C.:

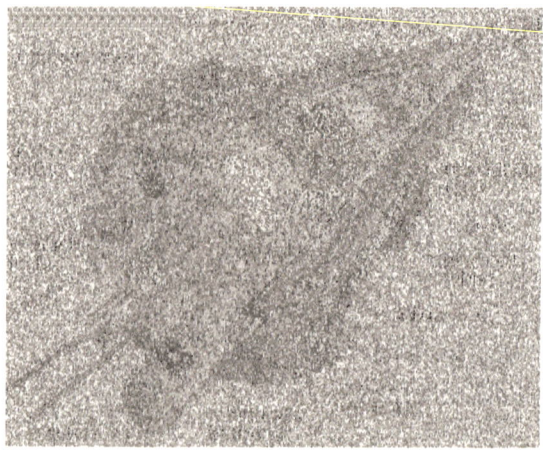

Figure 2.2: Dipole magnet in L.H.C.
(Credit: CERN)

The electromagnets are built from coils of special electric cable that operates in a superconducting state, efficiently conducting electricity with negligible resistance or negligible loss of energy. The temperatures of such superconducting magnets are maintained at -271.3 centigrade.

Talking about L.H.C. magnets, let me tell you that, there are thousands of magnets of different varieties and sizes which are being used to direct the beams around the accelerator. These include 1232 dipole magnets 15m in length and 392 quadrupole magnets, each 5m-7m long.

Just before collisions, another type of magnet is used to squeeze the particle closer together to increase the probability of collisions.

STAGE 5: COLLISION AND DETECTION

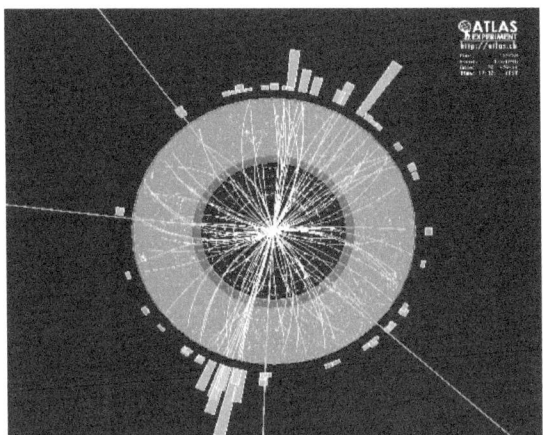

Figure 2.3: *ATLAS experiment particle collision*
(Credit: CERN)

The particles are so tiny that the task of making them collide is like firing two needles 10km apart with such precision that they meet halfway. The two adjacent tubes are made to intersect at four points where detectors are placed. Seven detectors have been constructed at L.H.C., located deep 100ft-300ft underground along with accelerating rings.

Two of them ATLAS experiment and Compact Muon Solenoid (CMS) are large in sizes and also are general-purpose detectors. ALICE and LHCb have more specific roles and the last three TOTEM, MoEDAL and LHCf are very much smaller and are for very specialized research. The summaries of the main detectors are given below:

1. **ATLAS:** Used to look for signs of new physics, including the origins of mass and Extra Dimensions.

2. **CMS:** Hunt for Higgs Boson and look for clues to the nature of Dark Matter.

3. **ALICE:** Studying fluid form of matter called quark-gluon plasma that existed shortly after Big Bang.

4. **LHCb:** Equal amounts of matter and antimatters were created in the Big Bang. It will investigate what happened to the missing antimatter.

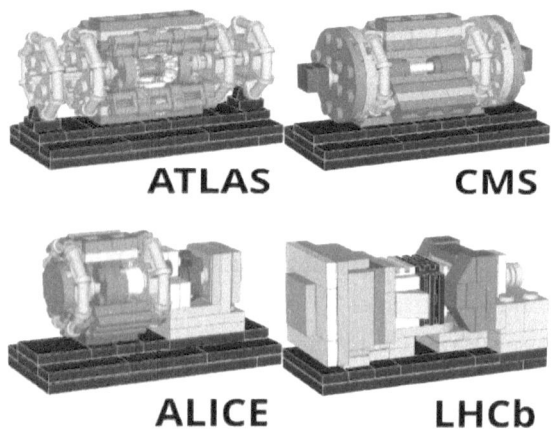

Figure 2.4: L.H.C. particle detectors
(*Credit: build-your-own-particle-detector.org*)

The L.H.C. at the moment is preparing itself for 20TeV of energy, unveiling physics beyond the Standard Model.

Standard Model of Particle Physics (SMP)

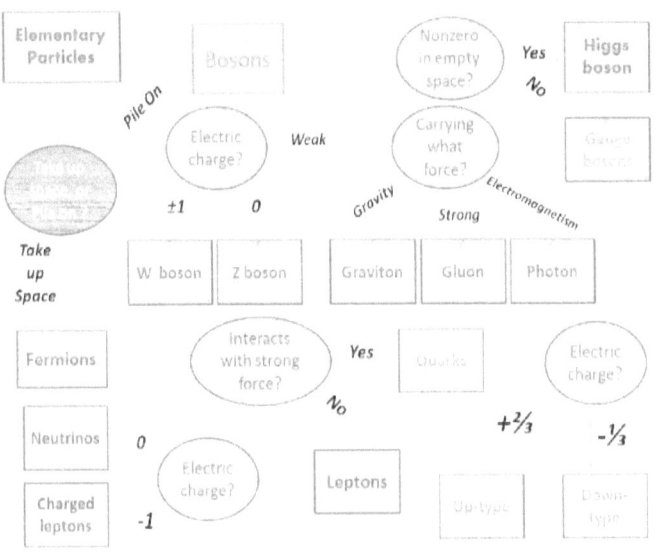

Figure 3.1: *Flow chart for the SMP*
(Credit: Wikipedia)

It's been more than a century since the discovery of the electron which is still thought to be a point-like particle. However, it's not the same as other particles like proton and neutron. It was found that protons and neutrons have a complex structure. The quest to find the ultimate building blocks of the universe had begun. From the past sixty years, the quest has led to discoveries of many particles, giving the zoo of particles. Theoretical and experimental physicists drove towards the development of the Standard Model which described three of the four fundamental forces in the universe and classifying all known elementary particles.

The term 'Standard Model' was coined by Abraham Pais and Sam Treiman in 1975, concerning electroweak theory with four quarks. In 1961, Sheldon Glashow combined electromagnetic force and weak nuclear force and developed electroweak theory (Chapter 6). In 1967, Steven Weinberg and Abdus Salam incorporated Higgs Mechanism (Chapter 6) into an electroweak theory which believed to give masses to W and Z bosons in electroweak theory and other elementary particles. Developments of particle accelerators led to the discoveries and experimental validations of many particles. After discovering Z Boson at CERN in 1973, the electroweak theory became widely accepted and Glashow, Salam, and Weinberg were honoured with 1979 Nobel Prize in Physics for the discovery. Later, in 2012, L.H.C. discovered the Higgs Boson which was the final missing piece for the confirmation of the Standard Model.

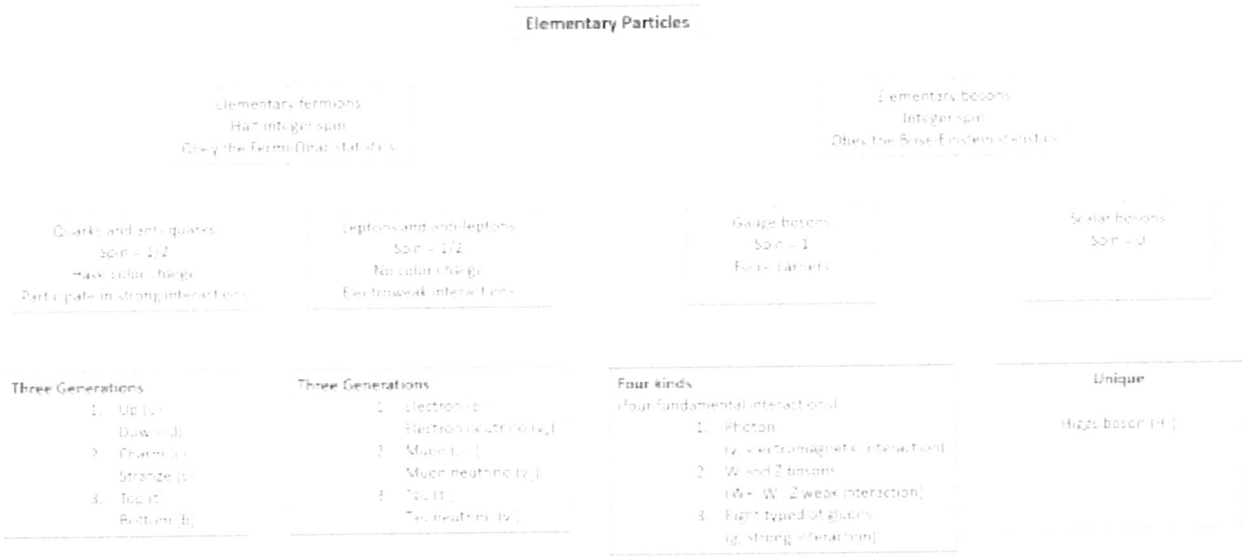

Elementary Particles

Figure 3.2: *Tree diagram of elementary particles in SMP*
(Credit: Wikipedia)

The Figure describes the classification of elementary particles in the Standard Model. The Standard Model is broadly classified as integer spin Bosons and half-integer spin Fermions which obey the Bose-Einstein statistics and Fermi-Dirac statistics respectively.

The Standard Model includes Fermions that consists of twelve particles and their corresponding anti-particles. Fermions which follow Pauli Exclusion Principle has six flavors of Quarks (up, down; charm, strange; top, bottom) and Leptons (electron, electron neutrino; muon, muon neutrino; tau, tau neutrino). Pairs from each classification are grouped to form a generation, which corresponding particles exhibit similar physical behaviour. These elementary particles have a spin with value ½ (particles) and -½ (anti-particles).

Gauge Bosons are force-carrying particles that mediate electromagnetic, weak and strong interactions. They do not follow the Pauli Exclusion Principle and have a spin with value 1 (vector boson). However, hypothetical particle 'graviton' which mediates the gravitational force, has a spin value 2 (tensor boson). The Higgs Boson has a spin with value 0 (scalar) and explains why particles (except photons and gluons) possess mass. Photon is a massless particle that mediates the electromagnetic force between electrically charged particles (eg. electrons, muon, tau). The W and Z Bosons mediates the weak interactions between particles of different flavors (Quarks and Leptons). The W^+, W^- and Z Bosons along with photons, collectively mediates the electroweak interactions. The eight massless gluons (Chapter 6) mediates the strong interactions between color charged particles (Quarks). The photons, W and Z Bosons, and gluons are well described by Quantum

Electrodynamics (QED), Quantum Flavordynamics (QFT) and Quantum Chromodynamics (QCD) respectively which will be discussed in Chapter 6.

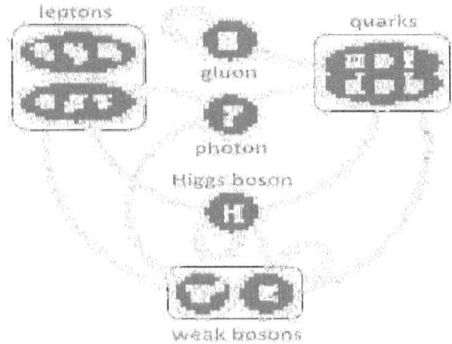

Figure 3.3: *Summary of interactions between particles described by SMP (Credit: Wikipedia)*

Figure 3.3 summarises the interactions between the particles described by the Standard Model. Unlike Fermions, Higgs, gluons and weak Bosons could also interact among themselves. In spite of this great theoretical work and experimental validation, the Standard Model falls short of being a complete theory of fundamental interactions. The major goal of physics is to find an ultimate theory of the universe i.e. the Theory of Everything (TOE) that would unite all theories and other known laws would be the special cases. The Standard Model is believed to be one of the prime candidates for the Theory of Everything (TOE). Therefore, physicists are attempting to discover particles beyond the Standard Model to answer the unanswered questions of the universe.

Beyond the Standard Model

As mentioned in the last chapter, the Standard Model is one of the prime candidates for the Theory of Everything (TOE). Though all particles in the Standard Model have been discovered, physicists believe that SMP is a small piece of a bigger puzzle which they called Beyond the Standard Model (BSM). BSM refers to the developments required to answer the unanswered questions by the SMP, few of which include, Why there's more matter than antimatter? What's the nature of dark matter and dark energy? From where does the SMP arise? Why gravity is so weak? And many more. BSM also refers to the theoretical and experimental developments required to resolve the deficiencies in the SMP which include, inconsistence with General Theory of Relativity, matter-antimatter asymmetry, failing to explain why stars at the outskirts of galaxies revolve faster than expected, failing to unify strong and electroweak interactions at high energies, and many other theoretical problems like hierarchy problem (Higgs mass through quantum corrections being larger than its actual mass), quantum triviality (inconsistent quantum field theory), strong CP problem (Chapter 6), etc.

Many models and theories have been proposed to resolve deficiencies in the SMP. In some of them, extra dimensions are introduced to explain why gravity is a much weaker force than other fundamental forces. In such theories, hypothetical force carrier particle called gravitons could be disappearing into these extra dimensions after being created in a proton-proton collision at L.H.C. There other theories that include new particles or interactions; common in many of these theories is the presence of new heavy bosons W' and Z'. A large theoretical framework of BSM that has been developed in the last decade is supersymmetry. Supersymmetry is one of the most discussed and studied extensions of SMP. The symmetry requires there to be a supersymmetric partner to all particles in SMP also called super-partners or supersymmetric particles or sparticles. Supersymmetry is expected to explain the hierarchy problem, nature of dark matter and unification of forces at high energies. In spite of having so much potential, there has been no experimental verification of sparticles. Due to the breaking of supersymmetry (a mathematical step to reconcile supersymmetry with actual experiments), the sparticles are much heavier than their ordinary counterparts; they are so heavy that existing particle colliders (including L.H.C.) may not be powerful enough to produce them.

Zoo of Particles

The following are the list of particles that have been discovered and hypothesized in the past sixty years. The list is just to give you a start to explore the zoo of particles. This is not an entire list, however, I aim to give a gist of the particles in particle physics. Readers could read the following list for self-exploration.

1. PHOTON:-

Symbol: ϒ

Mass: 0 eV/c^2

Electric Charge: 0 e

Color Charge: 0

Spin: 1

Status: confirmed

2. GLUON:-

Symbol: g

Mass: 0 eV/c^2

Electric Charge: 0 e

Color Charge: Octet

Spin: 1

Status: confirmed

3. W-BOSON:-

Symbol: $W^{\pm}$

Mass: 80.379 GeV/c²

Electric Charge: ±1 e

Color Charge: 0

Spin: 1

Status: confirmed

4. Z-BOSON:-

Symbol: Z^0

Mass: 91.188 GeV/c²

Electric Charge: 0 e

Color Charge: 0

Spin: 1

Status: confirmed

5. GRAVITON:-

Symbol: G

Mass: 0 eV/c²

Electric Charge: 0 e

Color Charge: 0

Spin: 2

Status: hypothetical

6. UP-QUARK:-

Symbol: u

Mass: 2.2 MeV/c^2

Electric Charge: +2/3 e

Color Charge: Yes

Spin: 1/2

Status: confirmed

7. DOWN-QUARK:-

Symbol: d

Mass: 4.7 MeV/c^2

Electric Charge: -1/3 e

Color Charge: Yes

Spin: 1/2

Status: confirmed

8. CHARM-QUARK:-

Symbol: c

Mass: 1.275 GeV/c^2

Electric Charge: +2/3 e

Color Charge: Yes

Spin: 1/2

Status: confirmed

9. STRANGE-QUARK:-

Symbol: s

Mass: 95 MeV/c^2

Electric Charge: -1/3 e

Color Charge: Yes

Spin: 1/2

Status: confirmed

10. TOP-QUARK:-

Symbol: t

Mass: 173 GeV/c^2

Electric Charge: +2/3 e

Color Charge: Yes

Spin: 1/2

Status: confirmed

11. BOTTOM-QUARK:-

Symbol: b

Mass: 4.18 GeV/c^2

Electric Charge: -1/3 e

Color Charge: Yes

Spin: 1/2

Status: confirmed

12. ELECTRON:-

Symbol: e⁻

Mass: 0.51 MeV/c²

Electric Charge: -1 e

Color Charge: 0

Spin: 1/2

Status: confirmed

13. ELECTRON NEUTRINO:-

Symbol: ν_e

Mass: Approx. 0 eV/c²

Electric Charge: 0 e

Color Charge: 0

Spin: 1/2

Status: confirmed

14. MUON:-

Symbol: μ⁻

Mass: 105 MeV/c²

Electric Charge: -1 e

Color Charge: 0

Spin: 1/2

Status: confirmed

15. MUON NEUTRINO:-

Symbol: ν_μ

Mass: Approx. 0 eV/c^2

Electric Charge: 0 e

Color Charge: 0

Spin: 1/2

Status: confirmed

16. TAU:-

Symbol: τ^-

Mass: 1777 MeV/c^2

Electric Charge: -1 e

Color Charge: 0

Spin: 1/2

Status: confirmed

17. TAU NEUTRINO:-

Symbol: ν_τ

Mass: Approx. 0 eV/c^2

Electric Charge: 0 e

Color Charge: 0

Spin: 1/2

Status: confirmed

18. HIGGS BOSON:-

Symbol: H^0

Mass: 125.18 GeV/c^2

Electric Charge: 0 e

Color Charge: 0

Spin: 0

Status: confirmed

19. X-BOSON:-

Symbol: $X^{\pm}$

Mass: $\approx 10^{15}$ GeV/c^2

Electric Charge: ±4/3 e

Color Charge: triplet or anti-triplet

Spin: 1

Status: hypothetical

20. Y-BOSON:-

Symbol: $Y^{\pm}$

Mass: $\approx 10^{15}$ GeV/c^2

Electric Charge: ±1/3 e

Color Charge: triplet or anti-triplet

Spin: 1

Status: hypothetical

21. AXION:-

Symbol: A^0

Mass: 10^{-5} to 10^{-3} eV/c^2

Electric Charge: 0 e

Color Charge: 0

Spin: 0

Status: hypothetical

22. MAJORON:-

Symbol: J

Mass: unknown

Electric Charge: 0 e

Color Charge: 0

Spin: 0

Status: hypothetical

23. X17 PARTICLE:-

Symbol: X17

Mass: 16.7 MeV/c^2

Electric Charge: 0 e

Color Charge: unknown

Spin: unknown

Status: hypothetical

24. STERILE NEUTRINO:-

Symbol: unknown

Mass: unknown

Electric Charge: 0 e

Color Charge: 0

Spin: 1/2

Status: hypothetical

OTHER HYPOTHETICAL ELEMENTARY PARTICLES:-

25. PREON

26. DILATON

27. LEPTO-QUARK

28. W' and Z' BOSONS

29. TACHYON

30. PLANCK PARTICLE

Heading towards Fields

For many years scientists are debating on the wave-particle behaviour of light and other fundamental things which make our universe. Some experiment like the photoelectric effect shows that light should be a particle, but other experiments like Young's double-slit shows that light should be a wave. Presently, the best of the prediction says that we have electromagnetic field; disturbance in this field causes waves, while unit excitation in this field gives rise to a photon. This intuition is more of mathematics than just saying. Before getting our head into this, let's talk about spin. Every particle that comes from these unit excitations may or may not have the property of spin. Spin may take a zero or non-zero (integer or half-integer) value. You may think spin as a particle's axis of rotation, but ill-advised to think like that while talking in quantum mechanics. As I said it's more a mathematical thing rather just thinking. To understand this concept let's learn some basic first.

6.1. CLASSICAL FIELD THEORY

The theory predicts the interaction of classical fields like electromagnetic field and gravitational field, the two fundamental forces of nature, with matter through field equations. Following are some classical fields:

6.1.1. SCALAR FIELD

Scalar field pervades the space-time i.e. it has a value at every point in space and time but no direction. Every field which we would be talking about, all pervades the space-time.

So whenever you here field, just consider that it pervades the space-time, even if it's not mentioned. Since spin is the direction of the axis of rotation and scalars don't show directions, we say that the particle formed by the excitation in such field is spin-0 particle i.e. it does not has any spin or intrinsic angular momentum, technically.

6.1.2. VECTOR FIELD

Vector field which pervades space-time has both, number as well as direction. The particles formed by the excitations in this field are thus spin-1 particles. They can take values of +1,-1 or 0 (no net rotation). Electromagnetism which I was talking about in the prior paragraph is an example of the vector field (Maxwell's theory).

6.1.3. TENSOR FIELD

Tensors are a generalized version of scalars and vectors. It gives linear relations between scalars, vectors, and other tensors; for example- dot product, cross product, linear maps, etc. Tensors can be represented as an organized collection (array) of numerical values. The rank or order of a tensor depends on the number of indices to label a component of that array. Let's take an example to understand this point; consider a vector where 1, 3 and 2 are the array of the numerical values which could be generally represented as **A**. Therefore, A_x A_y and A_z are the components with indices x, y, and z. From this, we conclude that in vectors we need one index or directional indicator (eg. x) per component (A_x). Therefore, vectors are tensors of rank one. On the other hand, scalars are dimensionless, they have an array of numerical values that are independent of indices or directional indicators, thus, scalars are rank zero tensors. Similarly, rank two tensors will have two indices per component (eg. A_{xx}, A_{xy}, A_{yx}, and A_{yy}); note that the example for rank two is represented for a tensor in two dimensions, for tensor in three dimensions we would have nine components instead of four. The reason why mathematics tells us to represent a component with two indices is that it helps us to fully categorize the entire system. For example, forces inside a solid object which have an area vector in x, y and z directions and on each of these surfaces there might be a force that has a component in x, y, and z-direction. So to fully categorize all the possible forces on all possible surfaces, we need nine components; each component with two indices, where one index of that component represents the direction of the force and other indexes of that component represents the direction of the area. So now coming back to the tensor field, just like a tensor, the tensor field is a generalization of scalar and vector fields. Einstein's theory of general relativity is governed by such a field that determines the curvature of space-time in the fourth dimension. A tensor field can take up to two units of angular momentum, thus, particles excited from this field are spin-2 particles and has five possible spin states (-2, -1, 0, 1, 2).

6.1.4. SPINOR FIELD

Spinors were introduced in geometry by Elie Cartan in 1913. In 1920s physicists discovered that spinors are essential to describe the intrinsic angular momentum (spin) of the electron and other subatomic particles. Spinors are the square root of vectors, they denote half-integer values. The property which characterizes spinors from vectors and tensors is that it transforms linearly when Euclidean space is subjected to a slight rotation. For example, when you rotate a vector by 360^0 the value of that vector remains the same. Similarly, if you rotate cross product of vectors i.e. a tensor by 360^0 the direction at which it's pointing remains the same. But when you rotate a spinor by 360^0, you get a value which is negative of the original value i.e. it points towards the opposite direction. Thus, rotations in three-dimensional space have a very special property and such rotations can mathematically be converted into a more general form which truly are the square roots of these rotations. In quantum theory, it's possible to construct a field of these square roots. This field too has a magnitude and a direction at every point in space, but it does not behave as a vector, instead, it behaves as a square root of a vector. The particle generated from the excitation of this field is spin-1/2 particles. Therefore, the intrinsic angular momentum of such particles can be plus or minus 1/2.

6.2. QUANTUM FIELD THEORY (QFT)

Erwin Schrodinger in 1925 developed an equation that gave us the probability of finding an electron in an atom. This equation is called Schrodinger's equation and it gave a core idea of Quantum Mechanics. However, the equation was not relativistic; it was Paul Dirac who combined Quantum Mechanics and Relativity in 1928. Quantum Field Theory (QFT) is a combination of classical fields, quantum mechanics and special relativity. It's a theoretical framework for constructing quantum mechanical models for subatomic particles. The difference between a classical field and quantum field is that a classical field is just a number-valued (it varies smoothly) whereas a quantum field is quantized and is an operator-valued. QFT suggests that there are no particles and waves; instead, our entire universe is filled with different kinds of fields, for example- electron field, up quark field, neutrino field, electromagnetic field, charm quark field, etc. Disturbances in these fields give rise to waves of that particular field, whereas excitation in these fields gives rise to a particle. These fields interact with each other and they transfer energy and momentum during such interactions.

When a field is excited we denote that energy state as a particle and this particle then interact with Higgs Field via Higgs Mechanism and thus defines the mass of that particle. To understand QFT in a better way let's take an analogy (The Candy Machine Analogy). Consider a candy machine that is filled with many delicious candies and to get candy from the machine you have to insert a coin. The candy thus won't be able to come out of the machine until and unless you insert a coin. Now the thing to remember is that coins are conserved i.e. if you are having 10$ and inserted 2$ into the machine then you will have only 8$ and the remaining would be in the machine. However candies are not conserved, they can be made in the factory and can be filled in the machine directly. Now moving with these given conditions compare the candy machine with a field, a coin with energy and candy with a particle associated with that particular field. Thus, whenever you give energy to a field it will give you a particle associated with that particular field e.g. if you give energy to an electron field then you will be getting an electron; If you give energy to an electromagnetic field then you will be getting a photon and so on so forth. Now similar to our machine there are about eighteen such machines which behave in a similar pattern. These machines can exchange coins (energy) but not candies (particles) among each other.

The analogy looks very simple and you should have figuratively imagined the interactions between the fields. However, in reality, it's much more complicated than this simple analogy. One of the key ideas of QFT is to explain the behavior of sub-atomic particles and their interactions via a variety of force fields. The two examples of QFT are Quantum Electrodynamics (QED) (describing the interaction of electrically charged particles and the electromagnetic force) and Quantum Chromodynamics (QCD) (representing the interactions of quarks and the strong nuclear force). Unlike the above two developed theories (QED and QCD) other forces (gravitational and weak nuclear) are also predicted to be explained by QFT. Let's talk about them one by one:

6.2.1. QUANTUM ELECTRODYNAMICS (QED)

If anyone asked me what is the best prediction made by mankind, then my answer would be QED. QED tells us how electrically charged particles interact with the electromagnetic force. In technical terms, it can be described as a perturbation theory (approximate mathematical prediction) of the electromagnetic quantum vacuum. In classical physics, when two electrons come closer to each other they repel since electric fields of those two electrons push them away from each other, however, in quantum mechanics that's not the case.

As per QED, the electric field is quantized and thus instead of force field, an electric field is created by a series of individual and discrete photons. Therefore, the electrons exchange one or many virtual photons which cause the repulsion to take place.

To give an analogy, imagine you and your friend in deep space with a ball in your hand. Now if you throw that ball towards your friend, due to vacuum you will be thrown in opposite direction and when your friend catches the ball, due to the ball's momentum he too will be displaced causing an increase in space between you and your friend. Now compare you and your friend as two electrons and the ball as a virtual photon and you will get a clear picture.

The scientists who got the explicit credit for developing QED theory go to Richard Feynman, Erwin Schrodinger and Sin-Itiro Tomonaga who shared the Nobel Prise in 1965. Among these three extra-ordinary scientists Richard Feynman's approach is far easy to understand. He came up with a series of pictures called Feynman diagrams that stand in for equations which made the whole process very easy to understand. He called this theory, "The jewel of physics" for its extremely accurate predictions of quantities like the anomalous magnetic moment of the electron and the Lamb shift of energy levels of hydrogen. Near the end of his life, Richard P. Feynman gave a series of lectures on QED intended for the lay public. These lectures were transcribed and published as Feynman (1985), *"QED: The strange theory of light and matter"*, a classic non-mathematical exposition of QED from the point of view articulated below:

The key components of Feynman's presentation of QED are three basic actions.

- A photon goes from one place and time to another place and time.

- An electron goes from one place and time to another place and time.

- An electron emits or absorbs a photon at a certain place and time.

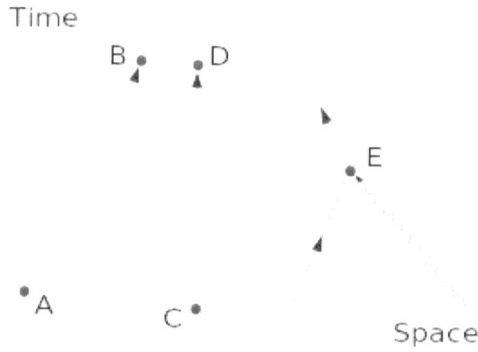

Figure 6.2.1: *Feynman diagram elements*
(Credit: Wikipedia)

These actions are represented in the form of visual shorthand by the three basic elements of Feynman diagrams: a wavy line for the photon, a straight line for the electron and a junction of two straight lines and a wavy one for a vertex representing emission or absorption of a photon by an electron. These can all be seen in the above diagram. It is important not to over-interpret these diagrams. Nothing is implied about how a particle gets from one point to another. The diagrams do not imply that the particles are moving in straight or curved lines. They do *not* imply that the particles are moving at constant speeds. The fact that the photon is often represented, by convention, by a wavy line and not a straight one does not imply that it is thought that it is more wavelike than is an electron. The images are just symbols to represent the actions above: photons and electrons do, somehow, move from point to point and electrons, somehow, emit and absorb photons. The theory does not explain how these things happen, but it does tell us the probabilities of these things happening in various situations.

6.2.1.1. FEYNMAN DIAGRAMS

Feynman developed pictorial ways that represent an equation for QED. His approach is thus easy to understand the basics of QED; however, solving the math is truly a different story and not everyone's piece of cake. A basic representation of a Feynman diagram is as follows:

Figure 6.2.2: *Feynman diagram representing the exchange of virtual photon between two electrons*

Figure 6.2.2 represents the exchange of a single virtual photon between two electrons. The straight line represents the path of electrons and the arrows indicate the direction of their motion. The wavy line represents the virtual photon and vertices or the point where straight and wavy lines meet is the point where the photon is emitted or absorbed. Interestingly this is not the only possibility in which an electron can transfer a photon. Other includes:

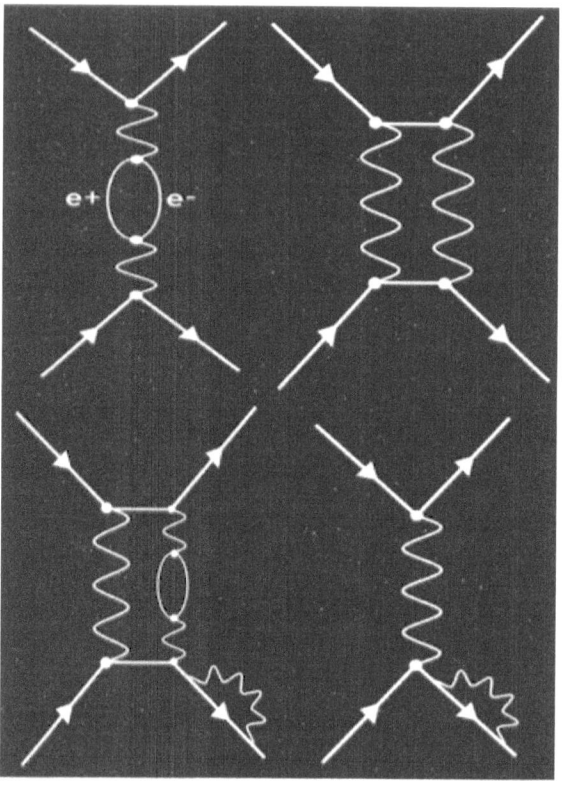

So how can we know which diagram is most likely to happen? We can find this by counting the number of vertices in the diagram. It turns out that photon emission or absorption is hard; specifically, each emission or absorption reduces the probability by about a hundredfold. Therefore, less the vertices more likely its outcome is. Since Figure 6.2.2 has a minimum of two vertices, it is the most probable outcome. Surprisingly, Figure 6.2.2 is not just a diagram but an equation. Below is a general idea:

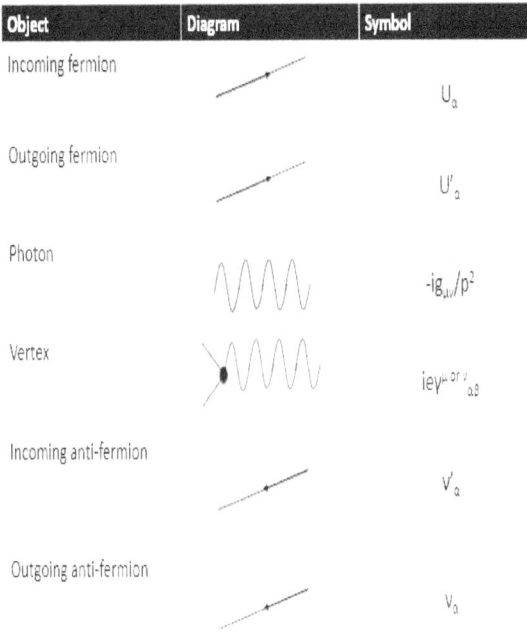

Object	Diagram	Symbol
Incoming fermion		U_α
Outgoing fermion		U'_α
Photon		$-ig_{\mu\nu}/p^2$
Vertex		$ie\gamma^{\mu \, or \, \nu}{}_{\alpha\beta}$
Incoming anti-fermion		V_α
Outgoing anti-fermion		V_α

Table 6.2.1: *Symbolic representation of Feynman diagrams*

Therefore by referring the table and figure, we conclude that the equation of Figure 6.2.2 is:

$$(\mathbf{u}_\alpha \, \mathbf{ie}\gamma^\mu{}_\alpha \mathbf{u'}_\alpha) \, [\mathbf{-ig}_{\mu\nu}/\mathbf{p}^2] \, (\mathbf{u}_\beta \, \mathbf{ie}\gamma^\nu{}_\beta \mathbf{u'}_\beta)$$

NOTE: The diagrams are drawn according to the Feynman rules, which depend upon the interaction Lagrangian.

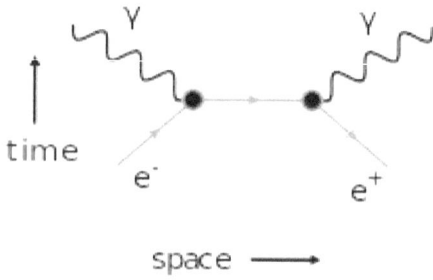

Figure 6.2.3: *Feynman diagram of electron and positron annihilation*

Let's consider another example. In Figure 6.2.3, there is an electron and positron annihilation (release of another particle after the collision between a particle and its anti-partner) interaction. The time arrow shows the direction of the event taking place as the time passes and space arrow represents the motion of the particles in space. Do note that there are two photons released during the interaction.

$$e^+ \, e^- \rightarrow 2\gamma$$

In the initial state (at the bottom; early time) there is one electron (e^-) and one positron (e^+) and in the final state (at the top; late time) there are two photons (γ). By using Feynman rules the equation of the above diagram is:

$$(\mathbf{u}_\alpha \, \mathbf{ie}\gamma^\mu{}_\alpha \mathbf{u'}_\beta \, \mathbf{ie}\gamma^\nu{}_\beta) \, [\mathbf{-ig}_{\mu\nu}/\mathbf{p}^2]^2$$

Similarly, there are other Feynman diagrams which you can go through like (Images source: Wikipedia):

➢ Beta Decay:

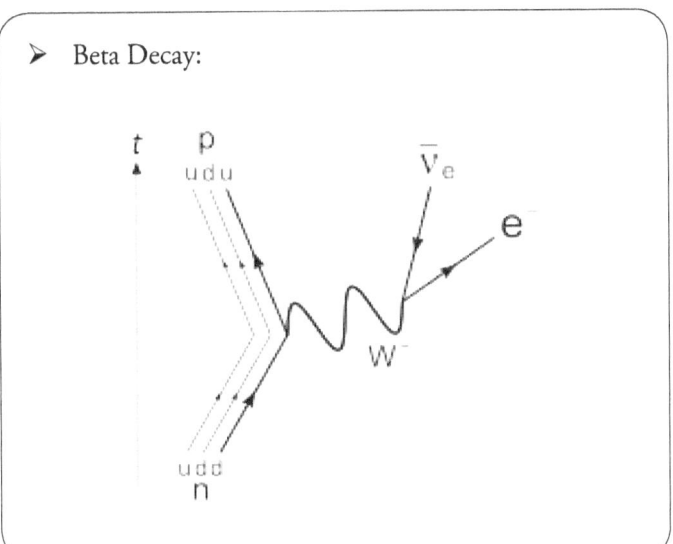

➢ Pair Creation and Annihilation:

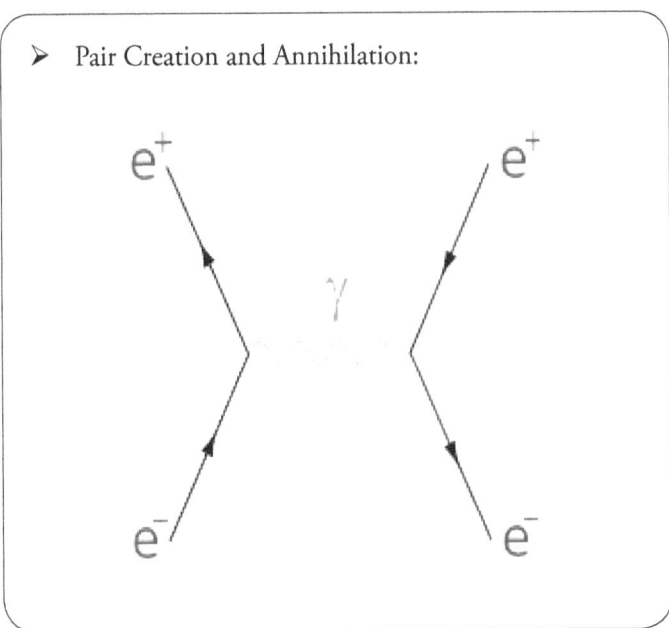

> Penguin Diagram:

> Higgs Boson Production:

A] Via gluon and top quarks:

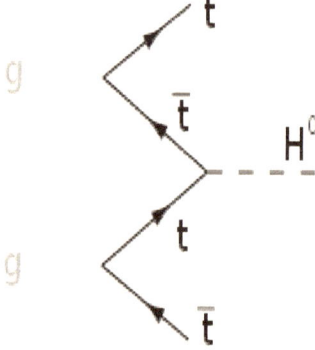

B] Via quarks and W or Z bosons:

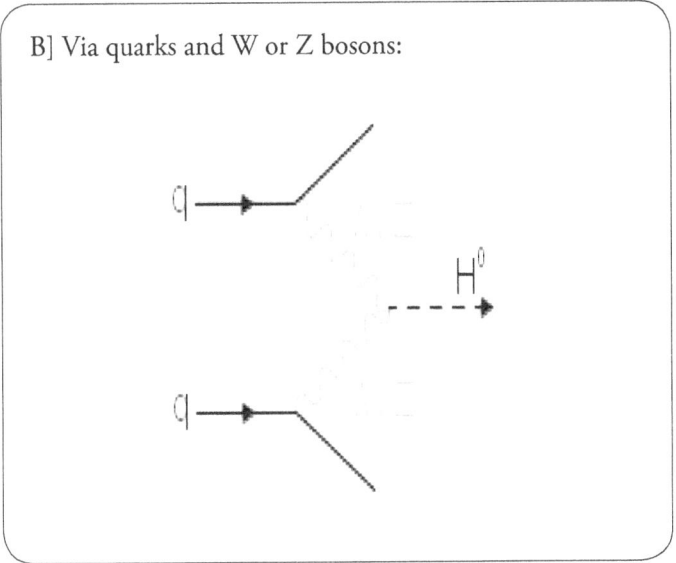

6.2.1.2. RENORMALIZATION

After multiplying all the vertices and propagators (eg. virtual photon) then comes the next step: The entire equation goes under an integral sign, integrating over all possible values of momentum for all internal lines. And this is where things can get nasty, as in most cases, this integral is divergent. Sometimes the divergence is genuine, representing a theory that behaves badly (e.g., naïve theories of quantum gravity). At other times, however, the divergence can be dealt with using various techniques of renormalization, e.g., so-called regulators that allow you to discard common (infinite) factors, leaving only the differences (which is what we are usually interested in).

6.2.2. QUANTUM CHORMODYNAMICS (QCD)

It's often said that a proton contains two up-quarks (one of the six 'flavors' of quarks) and one down quark and that a neutron as two down quark and one up-quark. However, that's not true. We know that there are six different 'flavors' of quarks: up, down, strange, charm, top and bottom; so if we define the electric charge of a proton as +1 then three of the quarks each have an electric charge of +2/3 and other three quarks have an electric charge of -1/3. Each quark has an associated anti-matter equivalent, called an 'anti-quark', containing the same mass but the opposite electric charge. In classical physics, the electric charge is responsible for the electromagnetic force. On the other hand, the strong nuclear force is due to a new phenomenon, which has no analogy in classical physics. This phenomenon is the 'color' charge. Quarks and anti-quarks can only take three possible values called red, blue, green and anti-red, anti-blue, anti-green respectively.

NOTE: The 'color' charges are the strong force charges and nothing to do with any specific color. It's just a fancy terminology given by the particle physicists to define strong charges.

Quantum chromodynamics (QCD) is a theoretical framework of strong interactions between quarks and gluons which make up composite particles like the proton, neutron, and pion (a meson). In QCD, gluons play a similar role as a photon in QED i.e. the gluon is the massless force carrier particle of strong nuclear force. However, unlike photon which is charge less, the gluons possess 'color' and 'anti-color' charges. Unlike quarks, gluons do not possess any electric charge, though they do represent 'color' and 'anti-color' charge. Each gluon possesses one 'color' charge and one 'anti-color' charge of a different type, such as red and anti-blue, red and anti-green, blue and anti-red, blue and anti-green, green and anti-red, green and anti-blue. In QED, the electromagnetic force is the result of virtual photons being exchanged between the particles having electric charge. Similarly, in QCD, the strong nuclear force is the result of gluons being exchanged between particles having a 'color' charge (in this case quarks). Therefore, it's the exchange of gluons that holds the quarks together inside protons and neutrons.

When a quark emits or absorbs a gluon, its flavor does not change i.e. an up-quark stays an up-quark and a down quark remains a down quark. However, the color of the quark changes. This is because the color is cancelled out by its associated anti-color, and the total amount of 'color' charge is always conserved, just as the total amount of 'electric' charge is always conserved. For example: when a gluon consisting of green and anti-red charges is absorbed by a quark with red charge, then, after absorption, the quark charge will be green. All composite particles, such as protons, neutrons, and pions are 'color-neutral' i.e. if we count the colors of all the quarks and gluons inside the composite particle, we will always find the amounts of red, green and blue to be exactly equal.

A gluon can split up into a quark and its associated 'anti-quark', provided that the total color charge is conserved (eg. A gluon consisting red and anti-blue charge will split into a quark with red charge and other quark with anti-blue charge). These events can also happen in reverse, with a quark and its associated anti-quark annihilating each other to produce a gluon. Therefore, neutrons and protons don't just each have three quarks, since many virtual 'quarks' and 'anti-quarks' are constantly being created and annihilated inside both neutrons and protons. Those three quarks in protons and neutrons are the 'valence' quarks. Protons and neutrons can emit a virtual composite particle consisting of one quark and anti-quark (called a meson), which can be absorbed by another proton or neutron. It's the exchange of these virtual composite particles, consisting of one quark and one anti-quark that creates the force binding protons to neutrons inside the nucleus of an atom. However, these virtual composite particles consisting of one quark and one anti-quark cannot exist for a very long time, which is why the strong nuclear force is a short-range force. Therefore, the protons and neutrons in a nucleus have to be very close together for this binding force to work and to overcome repulsion from the electromagnetic force between protons.

We know that photons are the force carriers between electrically charged particles. Also, photons do not themselves possess an electric charge and they will just pass through each other. On the other hand, gluons do possess a color charge, which means that unlike photons, gluons can absorb and emit gluons, allowing gluons to attract one another. Therefore, just like an electric flux, we have a gluon flux. However, unlike electromagnetic force which decreases as we increase the distance between charged particles, the strong nuclear force increases as we increase the distance between quarks. This consequence of the constant force between two color charges as they are separated is called color confinement, which is one of the two main properties exhibited by QCD. To increase the separation between two quarks within a hadron (eg. Proton, neutron, pion), ever-increasing amounts of energy are required. As we discussed, eventually this energy produces a quark-antiquark pair, turning the initial hadron into a pair of hadrons instead of producing an isolated 'color' charge. Although analytically unproven, color confinement is well established from lattice QCD calculations and decades of experiments. The other property of QCD is asymptotic freedom which is a steady reduction in the strength of interactions between quarks and gluons as the energy scale of those interactions increases (and the corresponding length scale decreases).

The dynamics of quarks and gluons are controlled by the quantum chromodynamics lagrangian. The techniques developed to work with QCD are lattice QCD, perturbative QCD, 1/N expression, Nambu-Jona-Lasinio model, etc. few of these techniques are briefly discussed below:

Lattice QCD: This non-perturbative approach uses a discrete set of spacetime points (called the lattice) to reduce the analytically intractable path integrals of the continuum theory to a very difficult numerical computation which is then carried out on supercomputers which were constructed for precisely this purpose.

Perturbative QCD: This approach is based on asymptotic freedom, which allows perturbation theory (set of approximation schemes) to be used accurately in experiments performed at very high energies.

1/ N expression: A well-known approximation scheme, the 1/N expansion, starts from the idea that the number of colors is infinite, and makes a series of corrections to account for the fact that it is not.

6.2.3. QUANTUM FLAVORDYNAMICS (QFD)

From the name, 'Flavordynamics' you might have guessed that QFD is some mathematical description of the dynamic behaviour of 'flavors' of quarks. This dynamic behaviour occurs during weak interactions eg. radioactive decay. Thus, QFD is a theoretical framework for weak nuclear interaction, describing the interactions between fermions (discussed in 'Standard Model of Particle Physics') and weak force carriers are known as W^+, W^- and Z bosons. Let's consider a top quark (one of the 'flavors' of quarks) and say it emits a photon (electromagnetic force-carrying particle) like in QED. The top quark does not change its identity ('flavor') before and after emitting the photon. Similarly, say the top quark emits a gluon (strong force-carrying particle) like in QCD. Here too the top quark does not change its identity ('flavor') before and after emitting the gluon. However, it does change its 'color' charge (as seen in QCD). Now say the top quark emits a W^+ or W^- or Z bosons (weak force-carrying particles). This time the top quark will change its identity ('flavor') after emitting the W^+ or W^- bosons, however, it will not change in case of the Z boson. This is because Z boson lacks electrical charge, thus, it won't be able to change the electrical charge of the top quark (as discussed in QCD top quark as an electrical charge of +2/3) causing the quark or the emitting particle to retain its identity. It's good to note, that these phenomena are equally valid when the force-carrying particles are absorbed.

6.2.3.1. VIOLATION OF CP-SYMMETRY

One of the most fundamental laws in physics is that the laws of physics follow symmetry. Just like conservation laws, symmetry has to be followed as well, otherwise, it will violate the foundations of physics. One of the laws of symmetry is C-symmetry and P-symmetry, where 'C' means charge conjugation and 'P' means parity transformation. C-symmetry implies that the laws of physics will be the same for matter and anti-matter particles whereas P-symmetry implies that the laws of physics will be the same for particles and mirror particles. The mirror particles are mirror images of the particles in the 'mirror world' which follows right-handedness. C-symmetry and P-symmetry hold well in electromagnetism, gravity, and strong nuclear interactions. However, in 1957, Dr. Chien-Shiung Wu experimentally observed that P-symmetry is violated in weak interactions. Her results shocked the physics community and there was a necessity to re-think the laws of physics. However, instead of restarting from scratch, physicists restored the P-symmetry by combing it with C-symmetry and making it a broader symmetry called CP-symmetry.

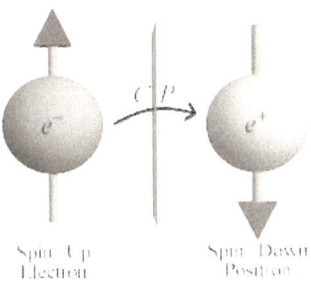

Figure 6.2.2: *CP-symmetry*

As CP-symmetry is a broader version of the law of symmetry, it couldn't violate but its constituent symmetry could be violated. However, in 1964 Val Fitch experimentally observed, that weak interactions also violated the CP-symmetry. Now two rules which physicists once thought to be fundamental laws of physics were broken. Therefore, they were left with the last set of theoretical defences i.e. to combine CP-symmetry with T-symmetry (time-reversal symmetry) and making it CPT-symmetry. Theoretical physicist restored the violation by CPT-symmetry. Therefore, C-symmetry, P-symmetry, and T-symmetry could be violated, however, CPT-symmetry is the real law of symmetry and has to be followed. To date there's no experimental evidence of the violation of CPT-symmetry, however, if it's violated, then many renowned theories like the Special Theory of Relativity and quantum field theory have to be re-written.

6.2.3.2. HIGGS MECHANISM

By 1960s physicists had built a strong understanding of strong nuclear force and the electromagnetic force. Attempts were being made to study weak nuclear force and to combine electromagnetism with weak nuclear force. This attempt is said to make a gauge-invariant theory with strong and electroweak force. However, the mathematics predicted two particles the photons (elementary particle of the electromagnetic force) and W, Z bosons (elementary particle of weak nuclear force) to be massless. We did know the photon is massless, however, experiments showed W and Z bosons are not massless bosons. That means the gauge-invariant theory is either wrong or incomplete. Other equations which explained the behaviour of fermions (electrons, quarks, and neutrinos) with the fundamental non-gravitational forces like strong and weak nuclear force and the electromagnetic force were very consistent and accurate, like the one below:

$$(- 1/4)F^{\mu\nu}F_{\mu\nu} - i\,\bar{\Psi}\,\eth\,\Psi$$

But when we tried to give mass by adding $\nu\,\bar{\Psi}\,\Psi$ to these matter particles ($i\,\bar{\Psi}\,\eth\,\Psi$) or even to force-carrying particles ($(- 1/4)F^{\mu\nu}F_{\mu\nu}$), the consistency or the symmetry of the equation was shattered away. Soon the physicist Peter Higgs and others introduced symmetric term $\phi\,\bar{\Psi}\,\Psi$ which didn't look like a mass term at first sight but when you add this term into the equation, you can see the beautiful symmetry of the equation to be maintained.

$$(- 1/4)F^{\mu\nu}F_{\mu\nu} - i\,\bar{\Psi}\,\eth\,\Psi + \phi\,\bar{\Psi}\,\Psi$$

The ϕ term can be re-written as $\phi = \nu + H$ and if we multiply $\bar{\Psi}\,\Psi$ on both sides we get:

$$\phi\,\bar{\Psi}\,\Psi = (\nu + H)\,\bar{\Psi}\,\Psi$$

$$\phi\,\bar{\Psi}\,\Psi = \nu\,\bar{\Psi}\,\Psi + H\,\bar{\Psi}\,\Psi$$

And $\nu\,\bar{\Psi}\,\Psi$ is the mass term and ν is the constant vacuum expectation value of the field that is permeating space and H is the scalar Higgs field. So this permeating field in this mathematical term gives rise to the mass of the particles.

This permeating field is the Higgs field which pervades the entire space. After the Big Bang, as the temperature started deducing the unified forces begin to separate. The electroweak force separated as electromagnetic force and weak nuclear force. Therefore, at a certain threshold temperature, the mathematical symmetry of electroweak force starts breaking down as it no longer exists as a single entity but as two separate forces. This is thus called 'symmetry breaking' and this breaking of the symmetry is produced by the Higgs Field and the mechanism which led to this symmetry breaking is called the Higgs mechanism. Higgs field is a scalar (spin-0) field and has a non-zero constant value in the vacuum (non-zero vacuum expectations), similar to an electromagnetic field. The Higgs field is pivotal in generating the masses of quarks and charged leptons (through Yukawa coupling) and the W and Z gauge bosons (through the Higgs mechanism). It is worth noting that the Higgs field does not "create" mass or responsible for the mass of all particles. Approximately 99% of the mass of baryons (proton and neutron), is due to quark-gluon binding energy (quantum chromodynamics binding energy), which is the sum of the kinetic energies of quarks and the energies of the massless gluons mediating the strong interaction inside the baryons.

In Higgs-based theories, the property of "mass" is a manifestation of potential energy transferred to fundamental particles when they interact ("couple") with the Higgs field, which had contained that mass in the form of energy. One of the common misconceptions with this field is provided with an analogy where resistance is shown to an object moving in a crowd. But in fact, the Higgs field does not work by resisting motion, it works on the coupling (interaction) with other quantum fields. There has been considerable scientific research on possible links between the Higgs field and the inflation – a hypothetical field suggested as the explanation for the exponential expansion of space within 10^{-35} of a second after the Big Bang (inflationary epoch). Some theories suggest that a fundamental scalar field might be responsible for this phenomenon; the Higgs field is such a field, and its existence has led to papers analysing whether it could also be the inflation responsible for this exponential expansion of the universe during the Big Bang.

6.2.4. QUANTUM GRAVITY (QG)

Quantum Gravity (QG), unlike any other field theories, is still a missing puzzle in the QFT domain. Theoretically, the theory is still under progress since combining General Relativity and Quantum Mechanics for gravity is extremely difficult. The current understanding of gravity is based on General Relativity which is within the framework of classical physics, unlike the other three fundamental forces we discussed above are described within the framework of Quantum Mechanics and Quantum Field Theory (QFT). While a quantum theory of gravity may be needed to reconcile General Relativity with the principles of Quantum Mechanics, difficulties arise when applying the usual prescriptions of Quantum Field Theory (QFT) to the force of gravity via graviton bosons. The problem is that the theory one gets in this way is not renormalizable (it predicts infinite values for some observable properties such as the mass of particles) and therefore cannot be used to make meaningful physical predictions. As a result, theorists have taken up more radical approaches to the problem of Quantum Gravity, the most popular approaches being String Theory and Loop Quantum Gravity. Although some quantum gravity theories, such as string theory, try to unify gravity with the other fundamental forces, others, such as Loop Quantum Gravity, make no such attempt; instead, they make an effort to quantize the gravitational field while it is kept separate from the other forces.

In the quantum realm, our experience of the universe appears in strange, but mathematically predictable ways. This math started with the Schrodinger equation which as we discussed (Introduction to QFT) tracks the probability waves through space and time. But the Schrodinger equation treats space and time in the separate old-fashioned way. However, it was Paul Dirac who came up with the relativistic wave equation and combined Special Relativity with Quantum Mechanics. QFT fully incorporates the melding of space and time predicted by Special Relativity, however, QFT still doesn't directly incorporate the warping of spacetime predicted by General Relativity. Therefore, it causes issues-some mild and fixable, others catastrophic, one of which is the Black Hole Information Paradox. The question of whether the information is truly lost in black holes has divided the theoretical physics community. In quantum mechanics, loss of information corresponds to the violation of vital property called unitarity, which has to do with the conservation of probability. However, Hawking radiation provides a part of the solution to the Information Paradox. It's clear that the information swallowed by Black Holes can be radiated back out into the universe via Hawking radiation. In a sense, both the source and the solution to the Information Paradox came from the discovery of Hawking radiation. Stephen Hawking derived this latter by finding a way to unite General Relativity and Quantum Field Theory (QFT). But that union was approximate and incomplete. This approach fails under strong gravitational effects on a smaller scale of space and time, like the central singularity of Black Hole or at the instant of Big Bang. Therefore, there is a necessity for Quantum Theory of Gravity i.e. Quantum Gravity.

6.2.4.1. STRING THEORY

String Theory (accounts for only bosons) is one of the radical mathematical approaches to renormalize quantum gravity. Instead of point particles, String Theory considers these particles to be strings vibrating in different modes (just like plucking a guitar string). String Theory was proposed while studying the strong nuclear force, however, it was abandoned by QCD. Later, it showed strong potential in developing the Quantum Theory of Gravity. And it did renormalize the Quantum Gravity, however, showed six spatial dimensions for the theory to work. In Superstring Theory (accounts for both fermions and bosons), it turned out that the unification of General Relativity and Supersymmetry known as Supergravity forms a part of hypothesized eleven spatial dimensional model known as M-Theory which is a generalized form of String Theory. Many scientists, to date, have not succeeded to renormalize Quantum Gravity in our three spatial dimensions. Therefore, there's still a long way to go before developing the Theory of Everything (TOE).

6.2.4.2. LOOP QUANTUM GRAVITY

Loop Quantum Gravity is a mathematical framework that considers the insights of General Relativity and represents the quantum behaviour of gravity. It's an attempt to combine Quantum Mechanics with General Theory of Relativity. Unlike General Relativity, in which spacetime is a continuum, the basic structure of spacetime in Loop Quantum Gravity turns out to be discrete. Therefore, they have a finite smallest (Planck's scale) value of length area, volume and time. One of the major challenges to combine Quantum Mechanics and General Relativity is that Quantum Mechanics displays the possible values of position and momentum on spatial coordinate system i.e. its background dependant whereas General Relativity curves spacetime, thus making it background-independent. Also, General relativity treats time as a different dimension, thus making it time-dependent, whereas Quantum Mechanics treats time to be independent. QFT also faces the same challenges as Quantum Mechanics while developing the Quantum Theory of Gravity. One such attempt was made by Wheeler and DeWitt who developed an equation describing the quantum evolution of geometry of space itself. However, it turned out that their equation was unsolvable. Theoretical physicist Carlo Rovelli has set on a mission to solve this equation by using Ashtekar's variables (representing geometric gravity) and spin network (quantum states which weave the spacetime itself). Once successful, it might be a breakthrough for the Quantum Theory of Gravity.

The Ultimate Theory of
the Universe

Physics has taken us a long way to understand the laws of nature. For centuries it has enhanced our knowledge to explore the universe. From Aristotle to Newton to Einstein, many scientists have contributed their life to the enhancement of mankind. And now we are hopefully at the end of our long quest to understand the universe. A theory that explains all physical aspects of the universe in a single theoretical framework of physics; an ultimate theory of everything.

In 1860, James Clerk Maxwell formulated an entire theory of electric field and magnetic field into a single framework called Maxwell's equations. His theory implied that electricity and magnetism are two facets of deeper and more fundamental phenomena called electromagnetism. In 1916, Einstein developed the General Theory of Relativity i.e. the field theory of gravity. Later, Einstein and others attempted to construct a unified field theory in which electromagnetism and gravity would emerge as different aspects of a single fundamental field. They failed, and to this day gravity remains beyond attempts at a unified field theory. During the Einsteinian era, very little was known about the nuclear forces. Development in quantum physics in the late 40s till the 70s led to the development of quantum field theory (Chapter 6) which described quantum behaviour of electromagnetism and strong and weak nuclear forces under the framework of Standard Model of Particle Physics (SMP (Chapter 3)). During this phase, particle physicists discovered basic building blocks of matter (quarks and leptons) which led to the enhancement of strong and weak nuclear forces.

In the 1940s, Quantum Electrodynamics (QED), the quantum field theory of electromagnetism, became fully developed (Chapter 6). At the same time that the picture of quarks and leptons began to crystallize, major advances led to the possibility of developing a unified theory. Theorists began to invoke the concept of local gauge invariance, which postulates symmetries of the basic field equations at each point in space and time. Both electromagnetism and general relativity already involved such symmetries, but the important step was the discovery that a gauge-invariant quantum field theory of the weak force had to include an additional interaction, namely, the electromagnetic interaction. Sheldon Glashow, Abdus Salam, and Steven Weinberg independently proposed a unified "electroweak" theory of these forces based on the exchange of four particles: the photon for electromagnetic interactions and two charged W particles and a neutral Z particle for weak interactions. It's this theory that led physicists to the discovery of Higgs Boson (Chapter 6).

The electroweak force was the first theory that successfully unified two fundamental forces (electromagnetic force and weak nuclear force). Attempts were made to unify strong nuclear force with electroweak force, however, it was an unsuccessful attempt until the development of Quantum Chromodynamics (QCD) during the 1970s (Chapter 6). Unlike gauge symmetry which unified electromagnetic force with weak nuclear force, it cannot unify strong nuclear force.

It turned out that new symmetry would be required to unify electroweak force with a strong nuclear force. This new symmetry was termed as supersymmetry. It was also found that forces have different strengths at different energy levels. It was found that with help of supersymmetry, the strength of the strong nuclear force and electroweak force will converge at high energies (around 10^{15} GeV which is a million million times as great as the energy scale of electroweak unification, which has already been verified by many experiments). Even the world's biggest particle accelerator (LHC (Chapter 2)) is incapable of producing such a high amount of energy. Therefore, experimental verification is a major challenge for such theories.

As I discussed in Chapter 6, it's difficult to develop a Quantum Theory of Gravity. As String Theory, Loop Quantum Gravity and other potential theories are still under development. Therefore, the unification of gravity with the other three fundamental forces is still a major challenge. It's estimated that the strength of gravity will converge with the other three fundamental forces at Planck's energy i.e. at 10^{19} GeV (Quadrillion times more energy than LHC). This unification of gravity with Grand Unified Theory (GUT) (unified theory of electroweak and strong nuclear forces) is termed as the Theory of Everything (TOE).

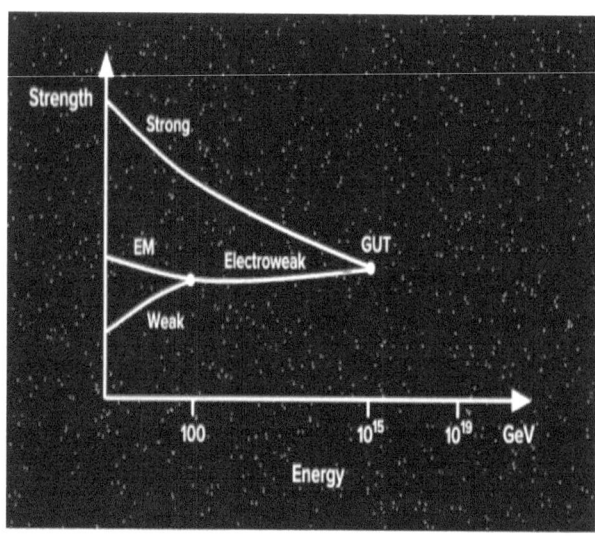

Figure 7.1: *Unification of fundamental forces at different energy levels*
(Credit: Fermilab)

To conclude, it would be difficult to construct a complete TOE all at one go. Therefore, physicists are making progress by finding and developing partial theories (String Theory, Quantum Gravity, Quantum Field Theory, etc.) Ultimately, one would hope to find a complete, consistent, unified theory that would include all these partial theories as approximations. To date, there's not a perfect model for GUT and TOE, however, scientists across the globe are optimistic about taking mankind to the ultimate triumph of human reason.

References

[1] David Griffiths. Introduction to Elementary Particles. Wiley, Sep 2008.

[2] CERN. Large Hadron Collider. https://home.cern/science/accelerators/large-hadron-collider.

[3] CERN. Facts and Figures about LHC. https://home.cern/resources/faqs/facts-and-figures-about-lhc.

[4] CERN. Experiments. https://home.cern/science/experiments.

[5] Fermilab. Particle Physics. https://www.fnal.gov/pub/science/particle-physics.

[6] Brain R. Martin, Graham Shaw. Particle Physics, 4th edition. Wiley, Jan 2017.

[7] Richard P. Feynman. QED: The Strange Theory of Light and Matter. Princeton University Press, Oct 2014.

[8] Fermilab. Videos by Don Lincoln. https://www.youtube.com/playlist?list=PLCfRa7MXBEsoJuAM8s6D8oKDPyBepBosS.

[9] PBS Space Time. Quantum Field Theory. https://www.youtube.com/playlist?list=PLsPUh22kYmNBpDZPejCHGzxyfgitj26w9.

[10] Stephen W. Hawking. The Theory of Everything, Special Edition. Jaico Publishing House, Sep 2006.